```
J                    438745
595.79                14.60
Abe
Abels
Killer bees
```

DATE DUE			

GREAT RIVER REGIONAL LIBRARY
St. Cloud, Minnesota   56301

KILLER BEES

BY
HARRIETTE ABELS

EDITED BY
Dr. Howard Schroeder
Professor in Reading and Language Arts
Dept. of Curriculum and Instruction
Mankato State University

PUBLISHED BY
CRESTWOOD HOUSE

LIBRARY OF CONGRESS CATALOGING IN PUBLICATION DATA

Abels, Harriette Sheffer.
  Killer Bees.

  (The Mystery of ——)
  SUMMARY: Describes the origin, behavior, and dangers of the aggressive African bees which are moving north from South America toward the United States.
   1. Brazilian honeybee—Juvenile literature. [1. Brazilian honeybee. 2. Bees] I. Schroeder, Howard. II. Title. III. Series.
  QL568.A6A24   1987   363.7'8   87-14085
  ISBN 0-89686-342-5

International Standard
Book Number:
0-89686-342-5

Library of Congress
Catalog Card Number:
87-14085

CREDITS

Illustrations:
Cover Photo: Dr. Orley Taylor
Dr. Orley Taylor: 5, 6, 9, 15, 27, 28, 36-37, 39
AP/Wide World Photos: 10-11, 21, 22, 30, 42-43
Bob Williams: 13,* 16-17, 24-25, 44-45
Oak Ridge National Laboratories: 19, 32, 40-41
Andy Schlabach: 34-35, 46
Graphic Design & Production:
Baker Street Productions, Ltd.

Copyright© 1987 by Crestwood House, Inc. All rights reserved. No part of this book may be reproduced in any form without written permission from the publisher, except for brief passages included in a review. Printed in the United States of America.

Box 3427, Mankato, MN, U.S.A. 56002

KILLER BEES

TABLE OF CONTENTS

Chapter 1 4
Chapter 2 8
Chapter 3 12
Chapter 4 20
Chapter 5 26
Chapter 6 31
Chapter 7 38
Map 46
Glossary/Index 47

438745

Chapter 1

"Killer Bees." The words sound like the title of a horror movie. But killer bees are more than images on a screen. These strange, angry insects are real . . . and they do kill!

It is not known exactly how many people have died so far by attacks of killer bees. Most of the attacks have taken place in rural areas in Latin America. But it is known that there have been hundreds of victims . . . and that killer bees are slowly moving north.

Killer bees were originally from Africa. They are slightly smaller than the European honeybees, which are common in the United States. Except for the way they act, their size is about the only physical difference between the two types of bees.

African bees get angry many times faster than European bees. They also sting more than eight times as much. When a European bee gets angry, it takes about four minutes to calm down. But an African killer bee takes as long as thirty minutes to quiet down!

All bees may get angry and sting their victims. But African bees seem to get angry in swarms. They have been known to continue biting their victim even after they have already stung them. After a bee stings some-

African bees like these have caused hundreds of deaths in Latin America.

Swarms of killer bees are slowly moving north.

one, the bee dies. But it can fly around sometimes for as long as an hour before it falls to the ground.

Entomologists (scientists who study insects) believe that all honeybees originally came from Africa. Centuries ago, some of them went to Europe on their own. They were not brought there by people. It is not known exactly why, but the European honeybee became a fairly tame insect. They have been raised as a business for hundreds of years. Well-trained bee handlers can even go into hives to remove the honey.

No one knows why the African killer bees get so angry. One scientist thinks it is because they had so many natural enemies. Badgers, ants, and even early man took the honey from the hives and destroyed everything else. The bees became fierce just to survive.

The swarms of killer bees are slowly moving north. There have already been several incidents of killer bee swarms in the United States. Many more are expected in the next ten years. Scientists are working very hard to come up with a solution to the problem. They are trying a number of different ways to prevent killer bees—and killer bee attacks—from spreading.

Chapter 2

The problem with killer bees started out with a minor accident.

Dr. Warwick Kerr is a geneticist, a scientist who studies inherited traits. He was the head of the Department of Genetics at the University of São Paulo in Brazil. In 1956, he won the Dreyfuss prize for his work in studying bee genetics. That was when he decided to go to Africa to bring back African honeybees.

Dr. Kerr wanted to develop a "super" honeybee, one that would work harder and produce more honey than the regular European honeybee common in North and South America. It was already known that African bees were more fierce than European or American bees, but Dr. Kerr was careful. When he brought back samples from Tanzania and South Africa, he chose queens that were supposedly the most gentle.

On the way home, there was an accident. In Lisbon, Portugal, a customs agent sprayed the bees with an insecticide called DDT. By the time they arrived in Brazil, all of the bees were dead.

Dr. Kerr was very upset. He chose his next batch of bees very quickly, and didn't screen out the fiercest ones.

Dr. Kerr brought the African bees to South America. He wanted to breed a "super bee," one that would be gentler and produce more honey than ordinary bees.

Forty-seven new African queens survived the trip. Dr. Kerr had planned to give the African queens to commercial beekeepers in Brazil. But once he was aware of the hateful nature of the bees he had imported, he decided to isolate the killer bee queens. He wanted to mate them with a more gentle type of bee, in order to improve them genetically. Dr. Kerr was sure he could produce a bee that was gentle as well as hard-working.

Somehow, another beekeeper visiting Dr. Kerr's laboratory let twenty-six swarms of Dr. Kerr's bees escape. Unfortunately, all of the swarms were led by

Once the African bees escaped, they spread rapidly. In the photo above, killer bees from different areas are compared.

an African bee queen. The swarms of killer bees flew into the wilds of São Paulo. It turned out that the genes of the African bees were so strong that they became the dominant genes when the African bees mated with Brazil's European honeybees. That meant that the fierce African bees were not becoming gentler—they were increasing!

Dr. Kerr has said that if he had it to do over, he would never have brought the African queens into Brazil. For years now, he has worked very hard to solve the killer bee problem.

Chapter 3

By 1962, it was obvious that killer bee swarms were spreading throughout Brazil. It was hard for people to understand what was happening. The honeybees that are native to Brazil are very gentle and seldom sting.

One of the first serious attacks was from a swarm of bees that had nested in a tile roof of a house near the airport at Rio de Janeiro. The swarm killed the dog that belonged to the owner of the house. A maid saw the attack and fainted. A neighbor called the fire department, which did not yet have a bee-fighting division. The firemen had butane torches on their truck, but decided not to use them. They were afraid of setting the house on fire. They had no pesticides with them, so they sprayed the bees using fire extinguishers filled with carbon dioxide gas. It didn't work. The bees became angry and turned on the firemen. Two of the firemen were stung so badly that they were sent to the hospital.

This same colony of bees had already killed fifteen hens on a farm, and had chased some workmen from a building that was being built. That time the firemen sprayed the bees with flaming gasoline. The bees flew off, but came back later. The second time, they were

In 1962, two Brazilian fire fighters were attacked by killer bees.

13

sprayed with an insecticide. That finally stopped the attack.

In September of 1965, a swarm of bees invaded a business area in downtown Rio de Janeiro. The swarm was so big that it looked like a black cloud. It hung over a building belonging to the armed forces. The people on the street panicked. Soldiers on duty at their machine-gun posts were driven out. This attack took place in the middle of the afternoon, with no warning.

One man found a beehive in the chimney of a local bar. The bees swarmed into the bar and stung more than five hundred people in the next three hours. When the bees finally left the bar, they attacked several nearby farms, killing chickens, dogs and horses.

It was thought that the killer bees' northern movement would stop when they reached the Amazon River. But that didn't happen. They crossed the river in 1971 and just kept right on going.

In 1972, there was a terrible killer bee attack on a small Brazilian farm. For some reason, a swarm of the bees became angry. They attacked several farm animals. Before long, two horses were dead, as well as three burros, five goats, and a dog. After that, the bees set out after the farmer and his family. The people ran into the house, where the farmer grabbed his shotgun. A man from a neighboring farm had heard all the noise and ran over to help. The poor farmer was so upset and confused that, when he fired the shotgun, it hit his neighbor. Fortunately, the neighbor was not killed, but

Killer bees enter a window at the front of a house.

A farmer's family takes cover as a "black cloud" of killer bees draws near.

16

he was badly injured.

The farmer and his family were stung many times, but they lived through the attack. When it was all over, though, the farm looked like it had been hit by germ warfare. There were dead animals lying all around. Their bodies were all distorted from the poisonous bee stings.

One year later, a cloud of the bees made an attack in a nearby town. They covered an old man. He died with over two thousand stingers in his body. They also attacked 130 other people in the town.

In that part of Brazil, killer bee attacks are so common that the fire department has a special bee squad. They get about four calls a day, usually when people find swarms nesting in abandoned cars, empty buildings or in trees.

The southern part of Brazil is hot and humid in the summertime. One hot, rainy day in December of 1973, the bees attacked a farmer. He was walking home from his fields where he had been harvesting corn. A swarm of bees had nested in the tile roof of the farm house. A dog was playing near the house and there were horses grazing in a nearby field. Suddenly, the horses reared up and began to race away with a swarm of bees chasing them. The bees attacked first in small groups, like waves of dive bombers. But before long, they attacked the poor animals the way they usually attack—all at once. Once the horses were dead, the bees turned on the dog. At this point the farmer ran to help him. The bees turned on the farmer, and within minutes he, too, was dead.

Afterwards, the farmer's wife, who had seen what was happening, said the blanket of bees attacked "like a wild bull." The bees were so upset that they continued to swarm and attack everything in sight for several hours. Finally, the fire department, using heavy smoke and strong insecticides, drove them off.

Dr. Kerr uses a "smoker" to calm African bees.

Chapter 4

In June, 1985, a colony of killer bees was found 140 miles (224 km) north of Los Angeles, California.

On his way to work one day, a worker at a California oil field saw a dead fox and a dead raven. He became curious, and later on in the day, during a coffee break, he decided to drive back to where he had seen the dead animals. As he came near, a rabbit jumped out of the brush. To the man's amazement, a swarm of bees rushed out of a hole in the ground.

The bees covered the rabbit until it couldn't be seen. In seconds, the animal was dead. Then the swarm rose into the air and headed straight for the vehicle the man was driving. They swarmed all over the vehicle, trying to get at the driver. It wasn't until he drove more than three hundred yards (270 m) away that the bees finally gave up and flew off.

It took two weeks before the animal control authorities arrived at the oil field to see what had happened. By then, most of the bees had disappeared. But when the officials dug down into the hole, they found a nest almost six feet (1.8 m) long and a foot (30 cm) wide, with three queen cells. This led them to believe that the bees had been nesting there for at least a year.

This empty comb was found near Los Hills, California, in 1985. It may have been the home of killer bees.

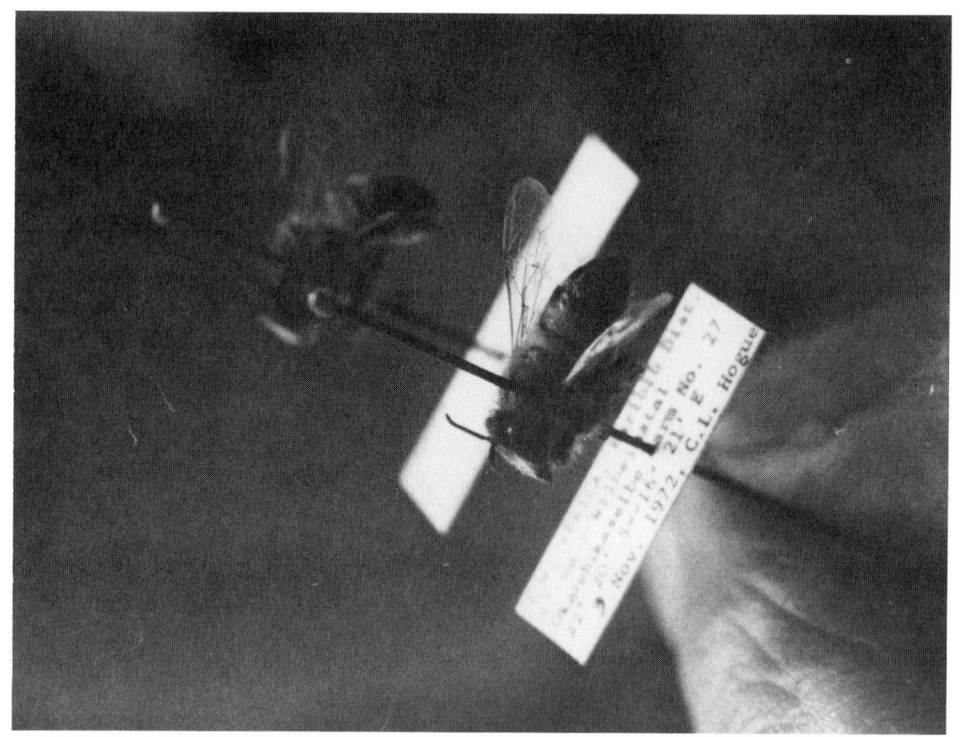

These were some of the first killer bees to enter the United States. They were found in Kern County, California, in 1985.

A few live bees had been trapped when they dug up the nest. These were sent to a laboratory of the California Department of Food and Agriculture. It took almost three weeks before the bees were examined. That was when they discovered that the oil field had been invaded by killer bees.

Entomologists from universities around the country quickly called a meeting. They set up a California State Advisory Panel on killer bees. The state authorities searched the land and buildings within ten miles (16 km) of the oil field where the colony was found. They are not sure how the bees got there. Some people believe that they came in a drilling pipe that was sent up from South America.

At first, it was suggested that aerial spraying be tried to get rid of the killer bees. But that idea was turned down because the bees live underground and the spraying probably wouldn't work. It was also decided not to destroy the local beekeepers' regular hives. Instead, the hives were carefully checked to make sure they had not been invaded by killer bees.

In their search, the bee experts found twelve colonies that had been invaded by the African bees. All of those colonies were destroyed. And in 1986, it was announced that the killer bees in California had been wiped out.

But the bee experts know that the next time it won't be so easy to get rid of the bees. In the California incident, there were too few killer bees to cause a serious problem. The next time it will probably be different. There will be so many African bees that the European bees won't have a chance. And the bee population will get meaner and meaner.

In this illustration, the African honeybee, or killer bee, is shown close up.

Chapter 5

All bees live in colonies or hives. When bees are raised as a business, there may be hundreds of colonies all together. Each colony can hold as many as sixty thousand bees. These hives are usually kept in wooden boxes. Bees that live in the wild build their hives in many places. They choose a hollow log or tree stump, or the chimney or roof of a house or barn.

The killer bees put a lot of their energy into producing more bees, instead of producing more honey, the way domestic bees do. They are very good at defending their hives or nests from their natural enemies.

The African bees fly much farther and faster than their European cousins. Although they are very good at gathering nectar, they aren't as choosey as the European bees. They also are not very good at telling the other members of the hive where they found the nectar.

A biologist at the Smithsonian Tropical Research Institute says that African bees seem to have a very good memory for odors. For instance, if the bees have attacked a rabbit that bothered them, they may attack any rabbit that wanders by a few days later. This biologist believes that may explain some of the stories of

African bees are very good at defending their hives from their natural enemies. This nest was found in Costa Rica in March, 1987.

All hives found in the wild should be considered dangerous.

"unprovoked" attacks. He also feels that the bees may be able to hear. It has been noted that they are often bothered by noisy vibrations, such as those coming from motorcycles.

All bees are attracted to objects that are dark. African bees are even more attracted to dark things. If they see a black ball, they will descend on it in an angry swarm.

The killer bees seem to be going through a population explosion. It is known that they have a higher reproductive rate than domestic bees. But the population increase is also probably due in part to their interbreeding with European queen bees.

Killer bees are most dangerous to people hiking or working in remote areas. If someone should knock over a hive or disturb the swarm in any way, the only thing to do is run as fast as they can. With European honeybees, entomologists tell us to either stand still or start moving away—slowly—if a bee gets angry. Not with killer bees! The best thing to do is to get inside of something, even if it is inside a car. If you get in and roll up the windows, the bees will forget about you and fly toward the light reflecting from the car windows.

Another thing to remember is that where there is one swarm of killer bees, there probably are others. People have to be aware that any beehive found in the wilderness may have already been taken over by killer bees. The best protection against them is to just stay away from all strange hives.

As killer bees spread into other countries, they often upset the commercial honey industry.

Chapter 6

Scientists have been mapping the movement of killer bees for the last twenty years. The bees have been spreading north at the rate of about two hundred miles (320 km) a year. Some scientists say that the bees cannot survive the cold weather if they go much farther north.

It was 1957 when the bees first escaped from São Paulo. They quickly spread in all directions. By 1963, they had moved into the city of Rio de Janeiro. Stories of attacks on humans and farm animals were already well known in that area.

By 1965, the bees had reached the northeast coast of Brazil. They moved over the mouth of the Amazon River by 1969. By 1971, they had crossed the river and were moving north.

They were also going in other directions. Some bees moved south into Argentina. They totally upset the honey industry there. The Argentinian government has talked about suing the government of Brazil.

Dr. Kerr says that one reason the bees are moving north faster than in other directions is that the climate is similar to that of the area in Africa from which the bees came.

It will probably take about five to eight years more

These researchers from the Oak Ridge National Laboratories are trying to develop a simple, hand-held device for finding African honeybees.

for the African bees to reach their northern-most point.

A professor from the University of Kansas, Dr. Orley Taylor, was given a grant by the United States Department of Agriculture to follow the northward movement of the bees. He spent time in both north and south Brazil as an observer. Dr. Taylor feels that amateur beekeepers are not equipped to handle the killer bees. He also feels that breeding the bees with a gentler species is not going to be the total answer.

Scientists from the Oak Ridge National Laboratories (ORNL) have been in Venezuela, testing an early warning system in case we have a massive invasion of killer bees. They spent a week in western Venezuela, where the African bees have become the main bee population. They wanted to find out if the sound made by the African bees is different from that of the European honeybees.

Scientists at ORNL have now developed equipment which, while not able to keep the bees out of the United States, will make it possible to quickly identify them.

The Department of Agriculture is trying very hard to find an answer to the problem, as entomologists expect major swarms of the bees to arrive near Brownsville, Texas, some time before the year 1990. The bees have already reached parts of Mexico and have caused the deaths of thousands of farm animals there.

Although there have already been several minor incidents of African bees in the United States, the large invasion is not expected until 1988-1989. It is expected

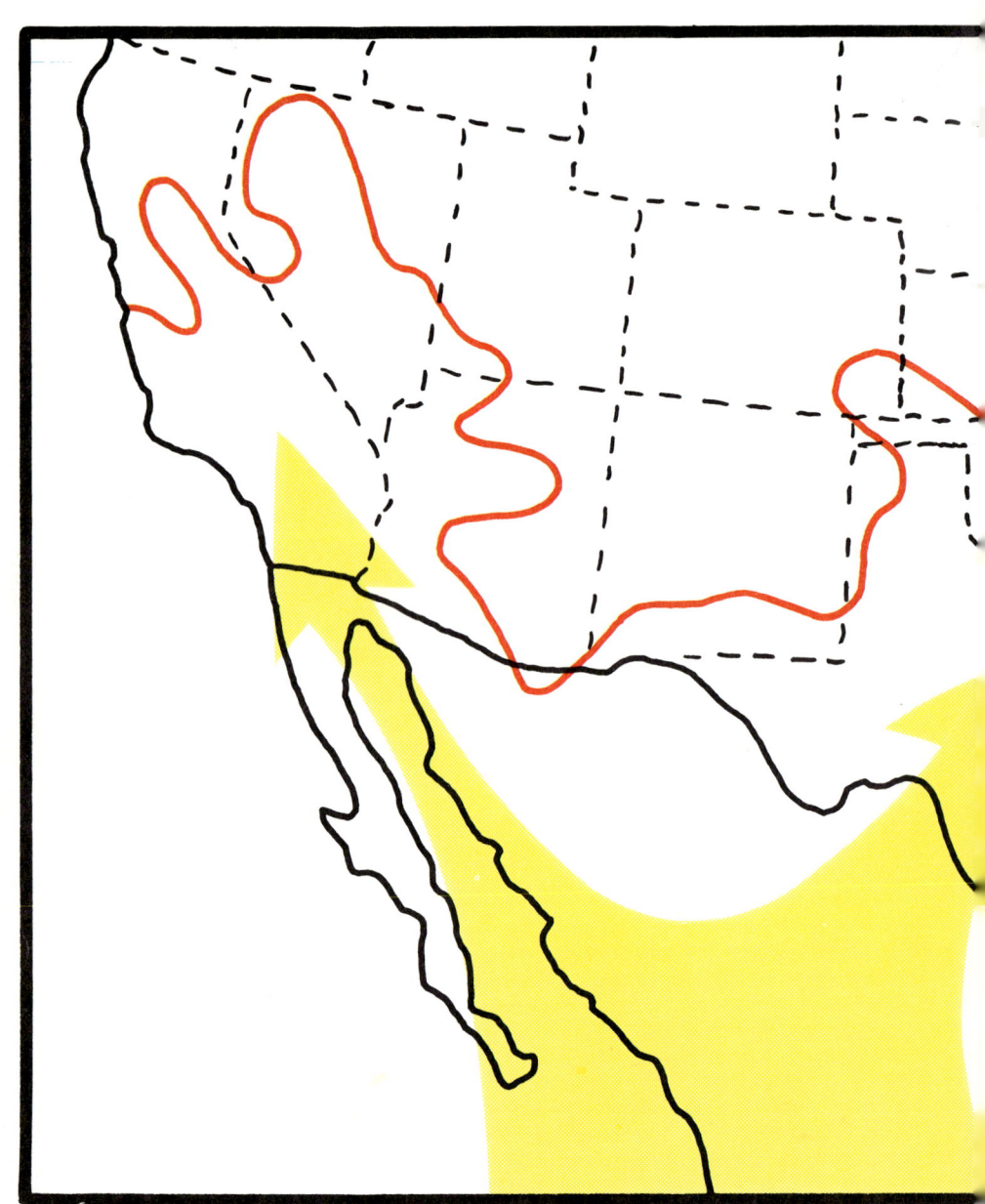

This map shows the likely path of the killer bees as they move toward the United States.

35

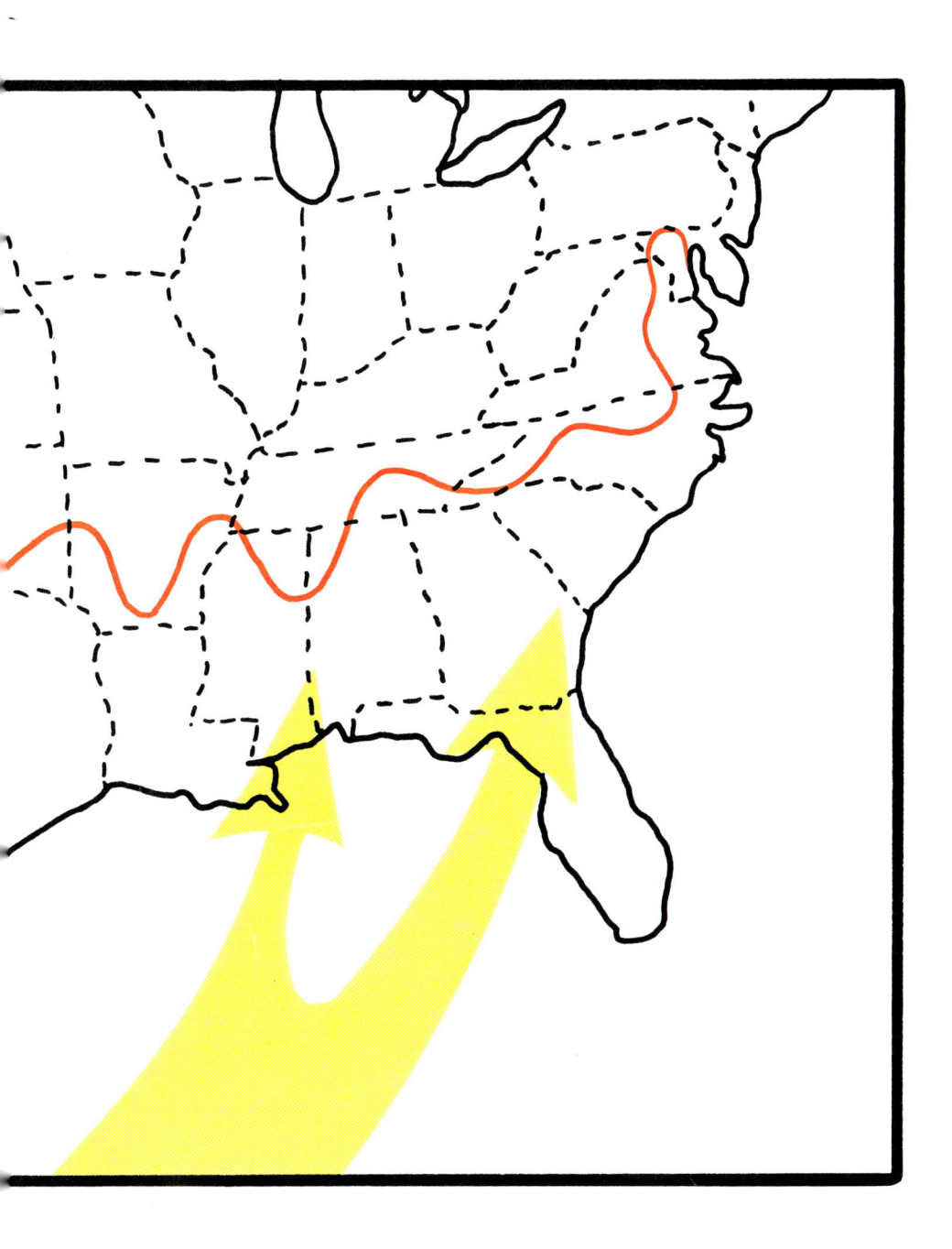

that they will first appear in large numbers around Brownsville, because the bees will come up from Mexico. It will probably take about two years for them to reach central and western Arizona. They may also spread into New Mexico and California.

The scientists think that the first few swarms will be small. We probably won't even know that they are here. These colonies will produce new colonies, and then begin to move on. In Texas, they will probably move

Beekeepers must carefully inspect their hives for African queen bees.

to the east. In Arizona, they will move north and west. The California bees will probably come from Baja, Mexico. Next they will move through the Imperial Valley, and then north along the coast.

The areas that the scientists worry about most are south central Texas, Louisiana (near New Orleans), southeastern Georgia, and most of Florida. Those are the areas where the climate is ideal and the amount of pollen and nectar available for the bees is very high.

Chapter 7

In the last few years, there have been many scientific studies done in the United States in order to solve the problem of the killer bees. Many scientists now believe that the problem will never be as serious here as it has been in South America. As the bees move northward, they become less dominant. This happens for several reasons. Since the bees originally lived in a warm climate (Africa), they probably will not be able to live through cold winters. It is possible that within five years after reaching the United States, they will have gone as far north as they can. Also, even after a commercial hive of European bees has been invaded by the African bees, it will be possible to replace the African queens with European queens.

Our government can also do many things to lessen the movement of African bees. If we have proper inspection techniques, it will be difficult for bees to "hitch a ride" from one state to another by nesting in empty pipes, or maybe large piles of wood.

The honeybee industry will have to make some changes when the killer bees arrive. The hives will have to be watched carefully to make sure they are not being invaded. Usually, in the spring, beekeepers re-populate

Scientists say that these bees may not be able to survive the cold winters in the mid-to-northern United States.

Studying these bees very closely may help us to prevent future tragedies.

41

their colonies with packaged bees from professional bee growers. It may be necessary to refuse shipments of all bees from areas where it is known that African bees have been breeding.

When the bees first spread to southern Brazil, the production of honey and beeswax dropped off. Today Brazil has more beekeepers, more managed colonies of bees, and more honey production than before the African bees arrived. Part of the reason was due to better

American beekeepers can learn much from the Brazilians' experience with killer bees.

beekeeping techniques. But also, the fierce behavior of the bees was controlled by crossbreeding them with a gentle species of Italian bees. This is quite easy to do in professional apiaries. Beekeepers in the United States can learn much from the Brazilian experience.

It may even be possible for us to avoid the terrible events that took place in South America when the African killer bees first arrived there.

If we are careful, we may be able to avoid the terrible killer bee attacks of South America.

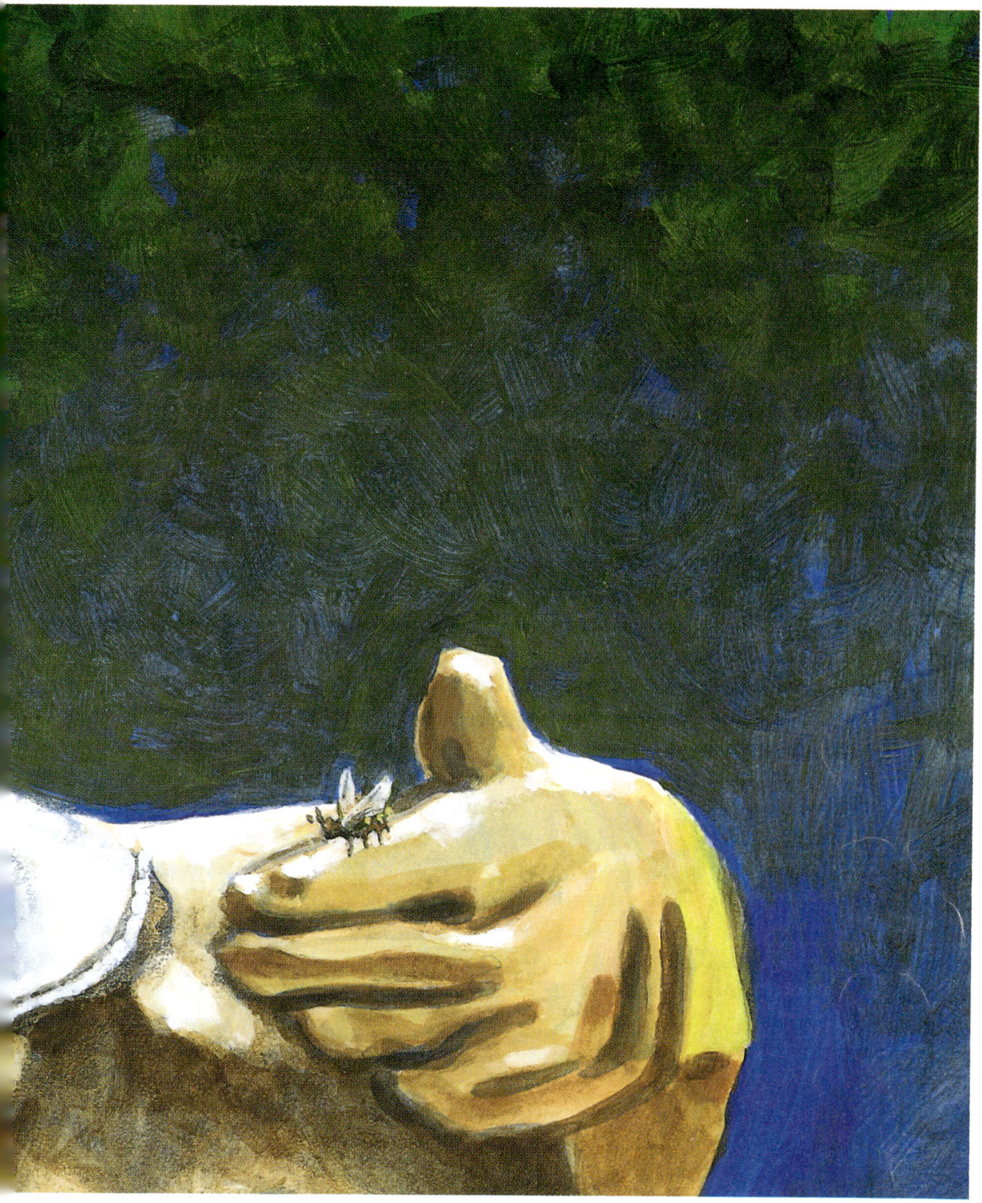

Map

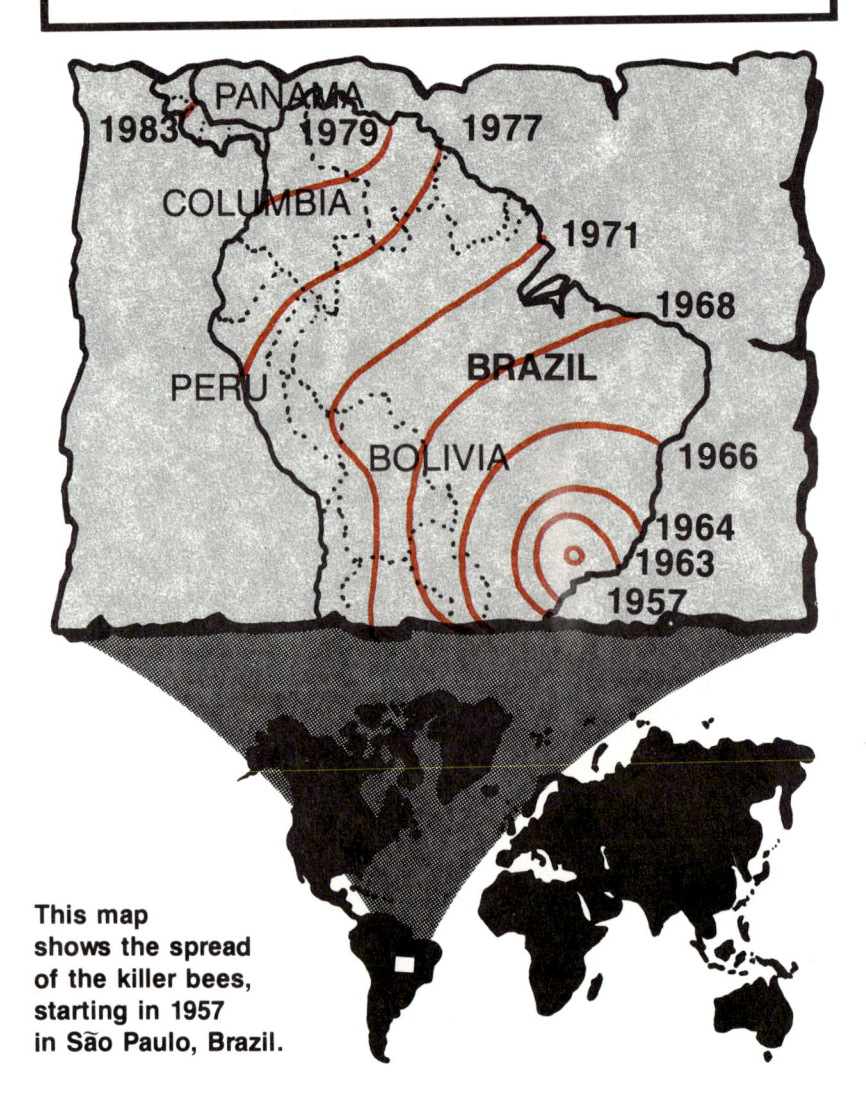

This map shows the spread of the killer bees, starting in 1957 in São Paulo, Brazil.

46

Glossary/Index

APIARY 43 — *A collection of hives of bees kept for their honey.*

BEESWAX 42 — *A waxy substance produced by honeybees, used to build their honeycombs.*

BIOLOGIST 26 — *A scientist who studies plants and animals.*

BUTANE 12 — *A flammable substance obtained from petroleum.*

CELL 20 — *The small, bordered "rooms" that make up a honeycomb.*

COMMERCIAL 10, 30, 38 — *Business conducted for profit.*

DISTORTED 18 — *Bent or twisted out of the normal shape.*

DOMESTIC 26, 29 — *Not wild; tame.*

DOMINANT 11, 38 — *More effective or superior.*

ENTOMOLOGIST 7, 23, 29, 33 — *A scientist who studies insects.*

GENES 11 — *A tiny part of a plant or animal cell that determines the characteristics that are passed down from parent to offspring.*

GENETICIST 8 — *A scientist who studies inherited tendencies.*

HIVE 23, 26, 27, 28, 29, 38 — *A structure in which bees live.*

INSECTICIDE 8, 14, 19 — *A substance used to kill insects, similar to a pesticide.*

NECTAR 26, 37 — *A sweet liquid in flowers. Bees use nectar to make honey.*

PESTICIDE 12 — *Poison used to destroy insects or rodents.*

POLLEN 37 — *A fine powder produced by flowers. Pollen makes it possible for seeds to form.*

QUEEN BEE 8, 10, 11, 20, 29, 38 — *A large female bee that lays eggs.*

SWARM 4, 7, 10, 12, 14, 18, 19, 20, 29, 36 — *A large group of bees that move together.*

47